Konan Emmanuel KOUADIO

Socio-economic integration of returnees to Côte d'Ivoire

Konan Emmanuel KOUADIO

Socio-economic integration of returnees to Côte d'Ivoire

ScienciaScripts

Imprint

Any brand names and product names mentioned in this book are subject to trademark, brand or patent protection and are trademarks or registered trademarks of their respective holders. The use of brand names, product names, common names, trade names, product descriptions etc. even without a particular marking in this work is in no way to be construed to mean that such names may be regarded as unrestricted in respect of trademark and brand protection legislation and could thus be used by anyone.

Cover image: www.ingimage.com

This book is a translation from the original published under ISBN 978-620-6-71146-9.

Publisher:
Sciencia Scripts
is a trademark of
Dodo Books Indian Ocean Ltd. and OmniScriptum S.R.L publishing group

120 High Road, East Finchley, London, N2 9ED, United Kingdom
Str. Armeneasca 28/1, office 1, Chisinau MD-2012, Republic of Moldova, Europe
Printed at: see last page
ISBN: 978-620-8-07777-8

Contents

Migration flows are not a new phenomenon, nor are they strictly modern (Bertrand BADIE, 1993, p 2). They are the consequence of various political, economic, social and humanitarian causes (Maurizio Ambrosini, 2010, p 3). In Africa in general, and in Côte d'Ivoire in particular, population migrations are caused by a variety of factors, including wars, socio-political unrest, natural disasters (droughts and floods) and the search for greater well-being. Geographical areas have become specialised as departure, transit and reception centres for migrants of all kinds. The centre of Côte d'Ivoire became a departure zone for able-bodied people heading for the forest areas and the country's major cities (M. Lessourd, 1985, p 85). But today, we are witnessing a return of migrants to this part of Côte d'Ivoire in general, and to the Boli sub-prefecture in particular. Generally speaking in Africa, and particularly in Côte d'Ivoire, the socio-economic integration of immigrants and returning migrants is a daily problem that needs to be resolved. This perilous equation hardly spares the populations and the authorities of their host towns, or even the national authorities. The young sub-prefecture of Boli, right in the centre of the Baoulé people, has seen the return of some of its vital forces who left their various villages since the start of the military and political crises in Côte d'Ivoire. In order to measure the extent of the socio-economic integration of returnees, we propose to carry out research into the ways in which returnees are integrated into the social and economic fabric of the Boli sub-prefecture.

GENERAL INTRODUCTION

The introduction to our work is structured as follows

1. Understanding the subject

Population migration is a global phenomenon that has existed for thousands of years. It affects both humans and animals. Côte d'Ivoire, which became a French colony in 1893, has not been immune. The populations of the centre of the country in general, and those of the Boli sub-prefecture in particular, have emigrated since the colonial period to the forested areas and urban zones of Côte d'Ivoire, and even to foreign countries. Although this sub-prefecture continues to see the emigration of its brave sons and daughters, it is now witnessing the immigration of certain Ivorian and non-Ivorian populations. These immigrants include a significant number of its "children".

The return movement to the region or village of origin has been neglected for a long time from a statistical point of view, which has made it difficult to study, and there is little research on it, particularly in the case of central Côte d'Ivoire. Return migration is a complex issue that can be analysed from several angles. Its close link with the economic and social development of the region of origin raises a number of issues. It depends on the migrants' profiles, the state of preparation for their return and the mobilisation of resources: financial capital and human and social capital.

Despite the massive arrival of new people in this administrative district, the perfect social cohesion between the inhabitants remains intact. What's more, economic activities are flourishing in all the localities of the Boli sub-prefecture. There is no sign of sadness on the faces of the inhabitants, whether they are non-emigrants, emigrants or returning migrants to the sub-prefecture.

The sub-prefecture of Boli has therefore become a place of population migration: many people have left for other, more advantageous places and many people have arrived in search of better political, security, social and economic conditions. Former emigrants are also returning to Boli without creating any real socio-economic problems.

The return migration of people to the Boli sub-prefecture and their socio-economic integration are therefore on the agenda.

What is the socio-economic integration of returnees in the Boli sub-prefecture? This is the focus of our study.

Through this study, we will show the following points:

> The profile of these returnees;
> The reasons for this return migration ;
> Ways of integrating into the host area.

2. Justification for the choice of subject

People's lives have always been shaped by population movements. These movements are characterised by the phenomenon of migration. People migrate in search of a better life. Throughout the history of mankind, migratory movements have been constant, affecting the whole world. These migrations have a variety of reasons: political, socio-economic and military. Since 1975, we have witnessed a different type of migration in Côte d'Ivoire. This is the return of certain migrants to their region or village of origin, or to an area in which they have lived for at least a year. This new migratory phenomenon seems to be on the increase as a result of the numerous economic and politico-military crises that have shaken the country since 2002. Located in the centre of Côte d'Ivoire, the young sub-prefecture of Boli has not remained on the sidelines of this phenomenon. The natives and people who used to live in Boli are returning for the reasons mentioned above.

These massive population movements have both positive and negative impacts on the development of the host site through the activities carried out by the migrants. From a socio-cultural point of view, the cohabitation between this returning population, with new cultures and knowledge acquired outside Boli, and the population that has remained in the area and is rooted in the local culture, is of interest to researchers.

From an economic point of view, the economic development of this administrative district that could result from this re-conquest migration, in view of the various economic activities, especially agricultural activities, could help the administrative, local, regional and even national authorities in their various development policies.

All these apparent multiple interests led us to focus on the socio-economic integration of returnees in the young sub-prefecture of Boli.

3. Presentation of the study area

Located in the centre of Côte d'Ivoire and crossed by the Abidjan-Ouagadougou railway line, Boli is the main sub-prefecture and can be reached by road (15 km from Didiévi and 82 km from Bouaké), and by rail (263 km from Abidjan, 61 from Bouaké). It is one of five sub-prefectures in the Didiévi department, namely Boli, Didiévi, Molonoublé, Raviart and Tié-N'diékro. Located 315 km from Abidjan, the sub-prefecture of Boli is part of the Bélier region. It is located at 4°48' 00 West and 7°13' 50.

North. The sub-prefecture is crossed from north to south by the railway linking Abidjan to Ouagadougou. It is bordered by the sub-prefectures of Molonou-Blé to the west, Didiévi to the south, Kouassi-Kouassikro to the east, and Raviart and Tié-N'diékro to the north.

Figure 1: Map of the study area

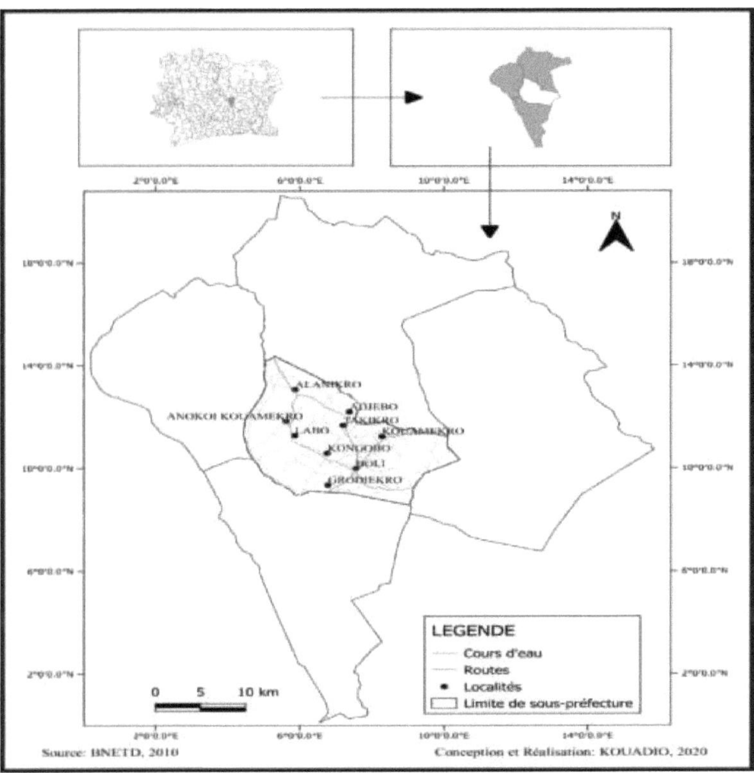

For decades, the Boli sub-prefecture has been emptied of its brave sons and daughters. As a result, it has been a departure zone for migrants. This migratory phenomenon, which began with the end of forced labour, continued in the 1980s, which were marked by economic and political crises. But around the year 2000, marked by land disputes and political and military crises, some of these people left to sell their talents elsewhere. The able-bodied and the brains are returning to the Boli sub-prefecture in search of socio-economic betterment and security in the lands of their birth or childhood. They flock from the forest regions and urban areas to settle in the administrative district whose capital is Boli. The population of the Boli sub-prefecture boomed from 1998-2014 (RGPH 2014). This has obvious socio-economic consequences, including social relations between the population and changes in the structure and well-being of the population.

Despite the end of the military crises, Boli continues to be populated by returning migrants. This flow includes the arrival of nationals and foreigners in search of better living conditions. These migrations have both positive and negative socio-economic impacts on the host environment in general and on the immigrant populations in particular. In fact, as the development of a community is linked to its society, it is unequivocally accepted that all the activities carried out by migrants have an impact on the life of the community, hence the need to study the impact of the movements of these return migrants on themselves through their mode of integration.

The Boli sub-prefecture is located in an area of high agricultural potential, and still has a high level of production of agricultural products, notably yams, cashew nuts, cassava, etc. In addition, this administrative subdivision has other assets that are no less significant, such as trade and services. These economic activities encourage intense population movements, so it is important to know the socio-economic integration patterns of returned migrants.
This study, entitled "The socio-economic integration of returnees to the Boli sub-prefecture", aims to use a scientific approach to analyse the socio-economic integration of returnees to the Boli sub-prefecture.

1 LITERATURE REVIEW

The literature on migration studies is abundant. It covers various aspects of the phenomenon, at international, regional and even national level. In Côte d'Ivoire, return migration is of prolific interest and various definitions of concepts arise from it. In order to deepen our understanding of the social and economic integration of migrants in Boli and to answer the many questions that arise from our subject, we feel it is necessary to define concepts related to our subject.

I.1 Clarifying concepts

Several concepts are important to define in order to understand our subject. These are : >

Insertion

The Larousse dictionary defines it as "the fact of being part of a group, the way of fitting in".

> **The return migration**

The concept of return migration is so broad that there is no universal definition. A number of definitions can be found in the literature. "Any migration to the place of origin, whatever the motivation" (J. P. Sanderson, 2015, p 56). The author defines return migration as a move to the place of birth or childhood. For the author, return migration is the return of an individual to the place where they spent their childhood hence in relation to their past. "A return migrant is therefore a person returning to his or her point of departure, without prejudging possible future migrations" (Véronique Petit, 2007, p 48). Véronique Petit defines a return migrant as an individual who returns to the place where he or she was before. J.L. Rallu defines return migration on the basis of the migrant's previous residence. All these authors refer to an individual's history to define return migration. MIREM (2007, p 2) defines return migration as "any person returning to the country of which he or she is a national, within the last ten years, after having been an international migrant (short or long term) in another country". "Ultimately, a returning migrant must meet the following criteria. They must spend at least one year in the country of origin after a stay abroad of more than one year" Marie- Laurence Flahaux (2009, p 13). Both authors take account of the sociological aspect by defining a return migrant as someone who returns to their place of origin. "A return migrant is defined as a person born in the country or region where he or she is registered or surveyed, having resided in another country or region at an earlier date" (J. L. Rallu, 2003, p 188). For the author, a return migrant is someone who returns to their place of birth after having resided elsewhere. "The return migrant is an individual who, having completed one or more migrations, returns to his or her sub-prefecture of birth" (C. Beauchemin, 2000, p 303). For Beauchemin, a return migrant is a person who returns to his or her place of birth after having moved to several other places. "Return migration is part of a return to rurality, to the countryside" (Thomsin, 2001, p 42). For the author, return migration is the movement of migrants from the city to rural areas. For these others, return migration is defined in terms of geographical location.

For us, a return migrant is someone who returns to live in the place of their origin, birth or childhood.

> **Socio-economic**

Socioeconomics or socio-economics is a blend of economics and sociology. It aims to integrate the tools of economics with those of sociology in order to examine the economic evolution of societies. According to the LAROUUSE dictionary, the term socio-economics is a qualifying adjective relating to social problems in their relation to economic problems. In other words, it refers to everything that concerns both the social and economic spheres and the relationship between them. The social refers to everything that concerns people in society. The interpretation of economic, political and social problems has led to the polysemy of terms which, according to the lexicon of the social sciences, initially belonged to only one of these social domains, which then corresponds to the confusion of the three. Economy, for its part, is defined in the common sense by the lexicon of the social sciences as savings, the art of managing household resources sparingly. The combination of these two terms refers to the way in which people interpret the economic, political and social problems inherent in their society. This relates to social problems as they relate to economic problems.

> Profile of returning migrants ;
> The reasons for migration ;
> The socio-economic integration of returning migrants.

I.2 Profile of returning migrants

8.   Mboup (2020, p 19) draws up a profile of the returnees in several paragraphs. Firstly, he states "a profile essentially characterised by masculinity and low levels of education" (B. Mboup, 2020, p 19). In his view, returnees are poorly educated and predominantly male.

This thesis of the minority of women among the B. Mboup returnees is supported by C. Beauchemin when he writes: "Women are less inclined than men to retire to the village". Mboup is supported by C. Beauchemin when he writes "Women are less inclined than men to retire to the village" (C. Beauchemin, 2000, p 121). For the author, this is explained by the fact that 'women do not enjoy the same social and material advantages in the village as men, hence their willingness to stay with their children. Next, B. Mboup (2020, p 19) adds that the age groups between 30 and 40 and 55 and 60 have the largest populations among the returnees. He adds that this pattern is proof that advanced age is a determining factor in return migration, and goes on to write that the majority of return migrants surveyed are married. Thus, married people are the most numerous among the returnees. He also adds that "cases of divorce and widowhood are found only among female migrants". This means that people who have become single only exist among women. Finally, Mboup (2020, p 24) sets out the results of his studies on the level of return migrants, stating that "at the end of this study, the migratory profile shows a large majority of migrants with a low level of training and education. The author states here that the respondents have a low level of education. Through "women often have a much lower level of education" Mboup (2020, p 22) indicates that female candidates for return migration are less educated than their male counterparts. The retired person thus appears to be the archetypal return migrant (Cris Beauchemin, 1999, p 3). For him, retirement seems to be a propitious time to migrate for several reasons: freedom from family and professional obligations, whether private or public. The chances of leaving the urban environment, initially highest for the elderly, in line with the classic image of the retired return migrant, increase over time for young people to the point where, at the end of the period, they have the highest rates of urban emigration (C. Beauchemin, 2004, p 178). "New forms of urban-rural migration (compressed, offloaded, etc.) have been added to the traditional returns (heirs, retirees)" (Cris Beauchemin, 2000, p 173). He backs up these arguments with a table that supports his hypothesis of retirement to the village. He supports this idea of 'retirement in the village' by thinking that 'Non-pensioned retirees are perhaps more inclined to return to rural areas where they can enjoy their possible investments (houses, plantations or businesses) and where the cost of living is lower' (Cris Beauchemin, 1999, p 3). Here, he argues that the return to the village is favourable to non-affluent retirees. The data show that return migrants are not mainly retirees and that many middle-aged adults migrate back. There appears to be considerable variability in return rates according to age, sex and level of education (J. L Rallu, 2003, p 188). This is supported by Beauchemin when he writes "return migration is far from being simply a migration of 'old people', retiring to the village" (Cris Beauchemin, 2000, p 118). He argues that the 15-30 age group is heavily involved in return migration. J. P. Sanderson (2015, p 57) takes a similar approach, pointing out that the work of Niedomyls and Amcoff on Sweden (2011) and Von Reichert (2001) on Montana highlights return migration affecting all ages of adult life through 'The work of Niedomyls and Amcoff on Sweden (2011) and Von Reichert (2001) on Montana highlights a phenomenon affecting all ages of adult life, as Jauhiainen (2009, p. 27) ".

I.3 The reasons behind this return migration

"The graph shows three main reasons for returning: family reasons (19.6%), investment projects (17.3%) and retirement (19%)" (B. Mboup (2020, p 19). For Mboup, family reasons are mainly the education of children, while the second and third reasons for return can be explained by certain socio-demographic and economic characteristics of these migrants and the circular nature of migration. The 2008 global economic crisis led to an unexpected rise in the number of immigrants returning home (OECD, 2017, p 268). Here, the European organisation points to the economic crisis as a factor that has encouraged return migration. "The region, the commune of origin then remains an anchor point, to which the family returns periodically, the place of identity continuity. In this case, return migration makes it possible to reconnect with dimensions of one's 'identity for oneself' that have only been temporarily put on hold, and which need to be restored on a long-term basis" Guichard-Claudic (2001, p 51). For Guichard-Claudic, the reason for emigrating is based on the desire to return to an area where one lived during childhood. According to Guichard-Claudic, the explanations put forward to understand return migration revolve around three key ideas: an economic explanation, a family explanation and an explanation linked to the life cycle. "The economic reason is as predominant in the interpretation of return migration as the explanation of the departure of migrants to their place of origin" (D. Potvin, 2006, p 30). Adult return migrants). For him, it is a question of their economic well-being, not forgetting the characteristics of the market and investment opportunities in the community of origin. He adds that the other reasons can be summed up as the degree or type of integration into the host environment, family ties and links with the environment of origin. According to D. Potvin, (2006, p 36) who draws on (Wymam, 2001; Waldorf, 1995; Dasgupta, 1981, Lipton, 1980, and Cerase, 1970), "the lack of work and job satisfaction would be largely responsible for migrants' integration failures and, as a last resort, their return to their home country or region". D. Potvin (2006, p 35) summarises the motivations for migrants' return in the writings of several authors (Wyman, 2001; Dasgupta, 1981; On-Jook and Kyong Dong, 1981; Cerase, 1970), who state that "integration, whether successful or not, appears to be an important factor in understanding the process influencing migrants' return to their place of origin". This is to say that we should not focus solely on failures to justify the return of migrants to their place of origin, because there are examples of successful return migrants. "The economic recession is not the cause of all forms of urban emigration, but is responsible for the reversal of flows between urban and rural areas" (Cris Beauchemin, 2000, p 173). For Beauchemin, falling incomes are a factor in the return of city dwellers to the village. He goes on to assert that the economic recession has precipitated the return of young people to the countryside, stating that 'the economic situation has played a selective role, particularly favouring the urban emigration of young people' (Cris Beauchemin, 2000, p 174). According to OTE/LA (1986, p 15), 42% of emigrants return earlier than expected for socio-economic reasons, 25% for family reasons, 15% for psycho-social reasons, 11% for favourable reasons and 8% for other reasons. But A. B. H. Zekri, (2007, p 15), based on surveys carried out in Tunisia in 1994, asserts the opposite of the Arab source, stating that favourable reasons, at 56%, are by far the main reasons for migrants returning to their countries of origin. This is followed by psycho-social reasons (17%), other reasons (11%) and family and socio-economic reasons (8% each). According to C. Beauchemin (2000), quoting Etienne (1970), Baoulé migrants have a sentimental attachment to their place of origin for three reasons. Firstly, it is a place of obligations (funerals, palavers to be settled, assistance to be provided, etc.), then a place of security in the event of health, social or economic problems, and finally a place where family status is affirmed and where professional experience is brought to bear. According to (P. Gubry et al, 1996, p 89), there are many reasons for returning to the village: Unemployment, lack of income, nostalgia and helping the family. E. Piguet (2013, p 156) summarises the different reasons for emigration when he states that it is rare for an individual to take the decision to emigrate because of a single factor. For him, the decision to emigrate is influenced by several factors, such as the socio-cultural context (the country, the religion, the religious majority or minority, the cultural-ethnic group, the cultural majority or minority), individual and psychological factors (gender, age, childhood context), the family (marital status, children, rank in the family, quality of family relationships, family support), quality of life (assessment of quality of life in the city, housing situation, environment, health system, etc.), networks (the family network, the family network, the family network, etc.).), networks (the family network, the network of acquaintances, migratory experience, transnational knowledge), economic factors (the economic level, satisfaction with resources, confidence in the economic future, employment prospects, entrepreneurship), the political context (satisfaction with institutions, the political system, human rights, equality between men and women) and the study context (study branches, success or failure in exams).

I.4 Ways of integrating into the host area

"The human capital, financial resources and social norms acquired by return migrants are an important source of development for many countries" (OECD, 2017, p 267). Here, the European organisation points to the economic crisis as a factor that has encouraged return migration. Drawing on Bovenkerk (1974), D. Potvin (2006, p 53) argues that "the effect of change in the environment will depend on the number of return migrants". For the author, a high number of return migrants would provide a critical mass of needs that would have an effect of change on the environment. "When they settle in rural areas, most urban emigrants naturally turn to agriculture (C. Beauchemin, 2004, p 189)". For Beauchemin, agriculture offers returnees an opportunity for economic integration. He backs up his idea by basing himself on the EIMU surveys, which show that 30% of returnees supplement agriculture with a non-agricultural activity (crafts, trade) when they arrive in rural areas (C. Beauchemin, 2004, p 189). This also means that, in addition to farming, returnees take up other economic activities, namely crafts and trade. For C. Potvin (2006, p 53), who draws on Lindstrom (1996), contradicts Beauchemin by stating that "the two main areas of investment are agriculture and residential". He argues that it is agriculture and housing that offer returnees a chance of economic integration. "It emerges that, while being vectors for modernising agriculture and diversifying rural activities, urban emigrants are confronted with multiple blocking factors that contribute to creating or accentuating local or national tensions" (C.

Beauchemin, 2004, p 189). Here, the author shows that returning migrants have difficulty integrating into their new community. This is not conducive to social cohesion. "Most urban migrants arrive in rural areas out of spite, when they have not succeeded in staying in the city. In these conditions, they arrive with no means of support, and their productive capacity is greatly reduced" (C. Beauchemin, 2004, p 190). Beauchemin describes the difficulties faced by returning migrants. These relate to finance (inability to employ labour and invest), agricultural training and access to credit. Financing difficulties are also raised by B. Ndione et al (2006, p 23): "The other explanations put forward by promoters are due to financing problems. To this financing problem, B. Ndione et al add family burdens, which are a source of difficulty for reintegration projects for returning migrants. According to them, the profits generated by a commercial project of 3 million CFA francs are unable to meet the needs of a family of more than ten members. According to B. Mboup, (2020, p 24), returnees with a low level of training and education have a low capacity for reintegration. What's more, their income drops when they return and their family burdens are increased by the predominance of minor children in their households. The author even adds that the difficulty of reintegration is greater for those who attended Koranic or Arabic schools and for the illiterate. "The return of the migrant to his or her village of origin is not without its problems, insofar as the return has not always been prepared. It is easy to imagine that problems of accommodation, food, availability of arable land and resumption of economic activity will arise" (P. Gubry, al, 1996 p 85). For these authors, the respondents are confronted with problems of accommodation, food, access to land and finance, but they do not fail to mention their source, which is the lack of preparation for the return to the village of origin.

2 ISSUES

Migration and the issues surrounding it are a global problem. The geographical areas that are home to these dynamic movements take on different functions depending on the specific features and connotations of migration. As a result, some areas of the oecumene are more emigration zones, while others are immigration zones, and still others are transit zones, or have all three functions. There are two main categories of migrant in the world: on the one hand, international migrants have increased in number. According to the United Nations, "Today, there are more people living in a country other than the one in which they were born than ever before. In 2019, the number of migrants in the world was approximately 272 million people, 51 million more than in 2010". Secondly, there is internal migration, which occurs in every country in the world. Statistics show that the very large waves of migration have recently diminished, in favour of a trend towards selective immigration. The characteristics of the current migratory phenomenon are the diversification of the countries of origin and destination, as well as the forms taken by migration. "Between a third and a half of migrants return home once they have achieved their objectives" (C. Daum, 2007, p1). So today we are witnessing the logic of reconstituting migratory spaces. More than 214 million people now live outside their country of origin for a variety of reasons: conflict, natural disaster, environmental degradation, political persecution, poverty, discrimination, inability to obtain basic services or the search for new prospects, particularly in terms of work and education (IOM, 2014). A larger number then live outside their village or region of origin for the same reasons. For Brouwez, (2017, p 2), there are four main causes that encourage immigration: causes linked to conflicts or political factors, economic causes, environmental causes and socio-cultural causes. Labour migration has become a means of subsistence for many families. In this case, the host area is seen more as a place of work than a place of residence (Fall, 2003, p 21). For an individual, the decision to migrate is often the result of a family strategy to maximise income (Amassari, 2004). Estimates made for OECD Europeans using the "labour force" surveys over the period 1992-2005, and for the United States using the 2000 population census and the 2005 American Community Survey, indicate that between 20% and 50% of migrants leave to work in order to earn income. In the five years following their arrival, immigrants either leave for third countries or return to their country of origin. In the latter case, more and more immigrants are returning either to their village of origin, their region of origin or their country of origin. Retrospective data on migration in Côte d'Ivoire show an increase in return migration after the military-political crisis of 2002-2003. "Matrimonial reasons were cited as the reason for 12% of all urban-rural migration between 1988 and 1993" (C. Beauchemin, 2000, p 164). In Beauchemin's view, the towns lost some of their population to the countryside between 1988 and 1993. This emigration towards the place of origin also takes into account emigration from the countryside to other parts of the countryside. These return movements are not without consequences. On the one hand, they affect the reintegrated development of these areas and, on the other, they influence the way of life of the returning populations. What about our work site? What is the profile of return migrants in the Boli sub-prefecture? What are the causes of return migration to the Boli sub-prefecture? Does this migratory movement hinder the development of the Boli sub-prefecture? How do returning migrants fit into the Boli sub-prefecture? This study was therefore initiated to help analyse and understand the phenomenon of return migration in the Boli sub-prefecture.

3 OBJECTIVES AND HYPOTHESES

I.5 Objectives

3.1.1 General objective

The general aim of this work is to analyse the socio-economic integration of returnees in the Boli sub-prefecture.

3.1.2 Specific objectives

Specifically, this involves :

> Determine the profile of these returnees;

> Identify the reasons for their return;

> Analysing the ways in which returning migrants fit in.

I.6 Research hypotheses

3.2.1 General hypothesis

Our general hypothesis is that return migrants are men aged over 45 with little economic power who integrate into economic, social and cultural activities to ensure their retraining.

3.2.2 Specific assumptions

The assumptions are as follows:

Assumption 1:

> The majority of returnees are illiterate men aged over 45 who have failed to make the grade;

Assumption 2:

> Socio-economic reasons justify return migration to Boli ;

Assumption 3:

> Returnees integrate through economic, social and cultural activities.

4 METHODOLOGY

I.7 Observation units

The units of observation for our work are all the villages and neighbourhoods of the Boli sub-prefecture. The analysis at village and neighbourhood level focused on the density of return migrants, their age, their marital status, their number of children, their origin and their level of education. We also looked at their living conditions and economic activities. This analysis enabled us to identify the estimated number of returnees, their profile, the general state of their economic situation and their needs.

I.8 Study variables

These variables are indicators that guide the search for the data to be collected. We have measurement variables, also known as quantitative variables, and assessment or qualitative variables. These are indicators that enable us to verify our hypotheses. Our research raises many questions, the answers to which will be guided by the variables that can be observed and assessed through the field surveys. These variables are of several types, notably socio-demographic, socio-economic, those linked to economic activities and those linked to socio-economic integration.

4.2.1 Socio-demographic variables

These variables concern the structure of the returning migrant population in the Boli sub-prefecture. They allow us to identify the modes of integration of return migrants in the Boli sub-prefecture.

Qualitative variables	Quantitative variables
- Origin;	- Number of children ;
- Where you live ;	- The number of each type of child ;
- Age;	- The number of children in and out of school ;
- Sex;	
- Ethnicity ;	- The number of working children ;
- Nationality ;	- The number of dependent children ;
- Level of education ;	- The amount invested.
- Marital status ;	
- Origin of spouse or partner(s)	

spouse(s)

- Religion

- The reasons why children drop out of school and do not go to school.

4.2.2 Variables linked to the determinants of return migration

These are variables linked to the reasons for their return. These variables enable us to judge the reasons for the return migration of these populations to the Boli sub-prefecture.

Qualitative variables	Quantitative variables
- Activities carried out during emigration ;	- The number of regions or countries of residence ;
- The reasons for the departure of the sub-prefecture ;	- The date of departure for emigration ;
- Achievements thanks to emigration ;	- The date of return from emigration.
- Last region of residence ;	

| -   Reasons for leaving the previous residence ;
-    The reasons for Boli's return;
-    Reasons for relocating to Boli. | |

4.2.3 Variables linked to integration into the socio-economic fabric

These variables are linked to the modes of integration of the returnees. These variables allow us to identify the modes of integration of these migrants in the sub-prefecture of Boli.

Qualitative variables	Quantitative variables
- Where the children live; - Place(s) of residence of spouse(s) ; - Problems encountered during installation ; - Relations with neighbours ; - Community conflicts ;	- The time set aside; - The surface area of the space in which the activities are carried out ; - New projects.

- Acquiring space in which to do business ;
- Activities carried out
- Dispute resolution
- Problems encountered in carrying out activities ;
- Soil quality.

4.3 Data collection techniques

> Documentary research

Our documentary research focused on general and specific works (Cris Beauchemin, for example) on return migration throughout the world, followed by those relating to this phenomenon in Côte d'Ivoire. This essential phase gave us a clear idea of our research problem. We visited the TUJLoG library and research institutes (IRD, IGT, CNRA, etc.) TINS as well. This research based on research studies (dissertations and theses) was very useful in terms of the information obtained. They have substantial characteristics and have undergone a rigorous process before being made available to the public.

> Field surveys

After the pre-survey, which took place from 09 to 19 August 2020, we moved on to making contact. This phase of direct contact with our study environment and the stakeholders took place over 20 days (from 01 to 20 September 2020) in all the localities in the administrative district of Boli. It began with a request for authorisation from the sub-prefect of Boli, Dame YAO Adjoua Albertine. Once the sub-prefect had given her permission and we had a meeting with her, we went from village to village and from district to district in the chief town, to take stock and talk to the village chiefs and our target population. The interview consisted of discussions and the return migrants were asked to fill in the pre-established questionnaire. These interviews provided us with useful information for our research into the integration of returnees to Boli.

4.4 The data collection method

In the absence of an appropriate database on returnees to our study area, a canvassing survey using the proximity method proved necessary to facilitate the identification and location of our target population. This approach, based on the dynamics of interpersonal relationships, made it possible to identify migrants who, once contacted, acted as relays towards other returnees like themselves. The mutual acquaintance of returnees living in the same locality was a key factor in the success of this survey strategy. Thus, by using the proximity method, we were able to list all the returnees in the Boli sub-prefecture. The size of our sample was therefore determined by the sum of the returnees identified. This methodological approach is a combination of qualitative and quantitative approaches. The qualitative approach is suitable for both data collection and analysis. It involved using the tools, techniques and principles of the accelerated participatory research method, such as individual interviews. This method was used during discussions with the return migrants surveyed and with the local and administrative authorities on the profile, the factors favouring return and the conflicts that have involved these migrants, the integration of return migrants and the development of the various localities. Participatory observation and document consultation are also part of the methods used. The quantitative approach consisted of structured interviews (questionnaires).

4.5 Data collection tools

The tools used in our work are interview guides, questionnaires and a census.

4.5.1 The interview guide and questionnaire

They are two elements that enabled us to refute or confirm our research hypotheses. They are developed taking into account the key concepts that make up the specific objectives of our work and the variables. These are state variables (sex and age of returning migrants), behaviour variables (the period of departure and arrival of migrants, their destinations and working conditions, etc.). They also enabled us to identify thought or opinion variables (people's knowledge and opinions on social cohesion, both between returning migrants and between returning migrants and other populations).

4.5.2 Sampling

We carried out a census. It consisted of a complete one-off or exhaustive survey using the close-to-the-person method. This made it possible to interview the entire population. It made it possible to observe all the basic units of this population. We conducted our survey from 01 to 20 August 2020 in all the villages of the Boli sub-prefecture. During the course of our work, we interviewed all the returnees who were present when we visited (some were travelling at the time). We were thus able to interview 122 people, who make up our sample of returnees to the Boli sub-prefecture. They are listed in Table 1.

Table 1: Populations surveyed

Places of residence	RGPH 2014			SEPTEMBER SURVEYS 2020		
	Women	Men	Total	Women	Men	Total
Adjébo	317	307	624	1	8	9
AkiikoiiiinwkiO	835	641	1476	1	2	3
Allanikro	584	539	1123	3	12	15
Anokoi-Kouamékro	159	153	312	2	12	14
Boli and Kongobo	3 891	3 350	7241	14	32	46
Grodiékro	761	861	1622	15	9	24
Labo	291	283	574	3	5	8
Takikro	168	138	306	1	2	3

Grand total	7 006	6 272	13278	40	82	122

4.6 Data analysis and processing

Analysis tools are the means used to test hypotheses and understand the various relationships between the variables measured. The data analysis consisted of a preliminary stage of analysing the survey forms. We then entered the data and processed it statistically using an Excel spreadsheet to compare statistical parameters such as averages and gross margins. Systemic analysis was also used. As a prelude to data entry, the processing consisted of designing the database of information collected and coding it. The surveys carried out as part of this work are based on a census of returning migrants. There are several areas of residence, namely the villages and neighbourhoods of Boli. The methodology used consisted of :

> A census of living quarters and their classification by place of residence (neighbourhoods and villages).

> Individual, semi-directive interviews supplemented by short closed or open-ended questions. The various locations visited were chosen according to criteria relating to the boundaries of the Boli sub-prefecture. The issues addressed in the interviews concerned the size of the district, its socio-economic situation, and the integration of returning migrants. The qualitative and quantitative data collected were tabulated, coded, entered, processed and analysed using Excel spreadsheets and *SPSS* statistical software.

4.7 The difficulties of the survey

On the whole, our research went well. Nevertheless, as with other research projects, we were confronted with difficulties that can be summed up as problems with the availability of stakeholders, access to documentation and statistical data, administrative red tape, and the reluctance or refusal of some respondents. We tackled them by taking measures adapted to each of them.

> In terms of access to documentation and statistical data, the use of libraries and the Internet provided satisfactory support;

> In terms of administrative red tape, we had to wait a few days before getting the Sub-Prefect's agreement to visit the various localities in the district;

> Due to the unavailability of some respondents, we were obliged to make appointments late in the evening (as the actors were often in the field), so that we often returned to our base at night, travelling more than ten kilometres;

> In the face of reluctance or refusal, a number of explanations and the support of certain players removed any doubt that our research would not be used for political ends

CHAPTER I: RETURN MIGRATION WITH A HIGH PROPORTION OF MEN

AND ILLITERATE PEOPLE

1.1 Socio-demographic and socio-economic characteristics of the Boli sub-prefecture

1.1.1 Socio-demographic characteristics

1.1.1.1 The population

The history of the people of the Boli sub-prefecture is closely linked to that of the large Baoulé group, specifically that of the N'zikpli Canton from which it originates. These people are known as Gnandjikro-sud, or southern Gnandji.

1.1.1.2 The population

Beginning with the Baoulé migration in the XVIIIe century, settlement of the district accelerated with the erection of Boli as a sub-prefecture in 2010 with the creation of new services. The district is populated by 13278 souls including 995 foreigners according to TINS (RGPH 2014). The population of the Boli sub-prefecture is made up of Baoulé N'zikpli, to whom have been added non-native Baoulé who are not N'zikpli, Malinké, other Ivorian peoples and hundreds of West African foreigners. All these peoples live in perfect harmony, with the exception of a small crisis that pitted the Baoulé against the Malinké in 2005, but which was quickly quelled.

1.1.2 Socio-economic characteristics

1.1.2.1 The habitat

At Boli, the landscape reflects a varied range of habitats. Unenclosed family dwellings are by far the most numerous (see photo 1). Individual dwellings are only just beginning to appear, as Boli, the capital of the sub-prefecture established for this purpose in 2010, has yet to really take off as a town. Precarious housing still exists in certain villages and even certain districts of Boli (source: personal surveys).

Photo 1: A family home in Akakouamékro

Photo credit: Kouadio Konan Emmanuel, 2020

1.1.2.2 Economic activities

Agriculture is the main activity in the district. It is made up of food crops (yams, manioc, maize and rice) and export crops dominated by cashew nuts, which have been joined by oil palm and teak wood. Alongside agriculture, trade (thanks to the train station until 2011 and the only weekly market in Boli) is doing its bit. Cattle and goats are also reared in small numbers. It's worth noting that, as we passed through, timber harvesting for charcoal production was in full swing. Services are coming on stream, particularly in education, health and a number of public and private administrative services.

1.2 Migration profile

1.2.1 Place of residence

These are the places where we surveyed them. We counted 8 localities, namely the capital of the sub-prefecture (Boli) and 7 other villages that make up this administrative district. Boli is a large urban area with several districts, including Kongobo, a village that moved to Boli between 1998 and 2014. The population of each village in 2014 according to the RGPH is highlighted in Table No. 2.

Table 2: Populations surveyed

Place of residence	Population RGPH 2014	Population surveyed
Adjébo	624	9
Akakouamékro	1 476	3
Allanikro	1 123	15
Anokoi-Kouamékro	312	14
Boli	7 241	46
Grodiékro	1 622	24
Labo	574	8
Takikro	306	3
Grand total	13 278	122

Sources: Kouadio surveys, September 2020

1.2.2 Dominated by men and older people

The ages have been classified into brackets ranging from 18 to 60 and over. Thus, we have the following age groups: [18-30 years [; [30-45 years [; [45-60 years [; [60 years and over [. The younger age groups have fewer return migrants. In fact, the [18-30[age group has only 5.74% of the population among those surveyed. The [30-45] age group represents 21.31%, the [45-60] age group has 24.59% and finally the [60 and over] age group, with a proportion of 48.36%, has almost half of our target population. The breakdown of respondents by gender shows that around two out of three returnees are men (67.21%). Among the youngest respondents (aged 18-45), men accounted for 57.58%, while among the oldest (aged 45 and over), men accounted for 70.79%. Figure 1 illustrates the distribution of the return migrant population by gender and age.

Figure 1: Breakdown of the returnee population by gender and age

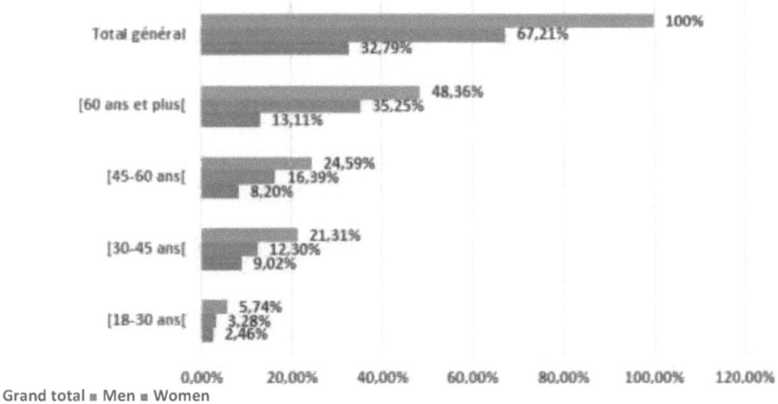

Sources: Kouadio surveys, September 2020

1.2.3 Couples dominated by common-law unions

Marital status describes the individual's marital status. They may be single, divorced, married, in a common-law relationship or widowed. The survey revealed the following statistics: single: 17.21%; divorced: 0.82%; married: 6.56%; in a common-law relationship: 60.66% and widowed: 14.75%. Unmarried couples are those whose civil marriage has not yet been formalised. According to our investigations, the vast majority of migrants have celebrated their customary marriage. Furthermore, here a divorced person is someone who was previously legally married and then divorced. We note a high proportion of respondents in common-law unions, and only one man is divorced. It should be added that no women are married and that 61.11% of widowers are women. Figure 2 gives an overview of marital status by gender. The vast majority of our target population is monogamous (77.86%).

Figure 2: Marital status by gender

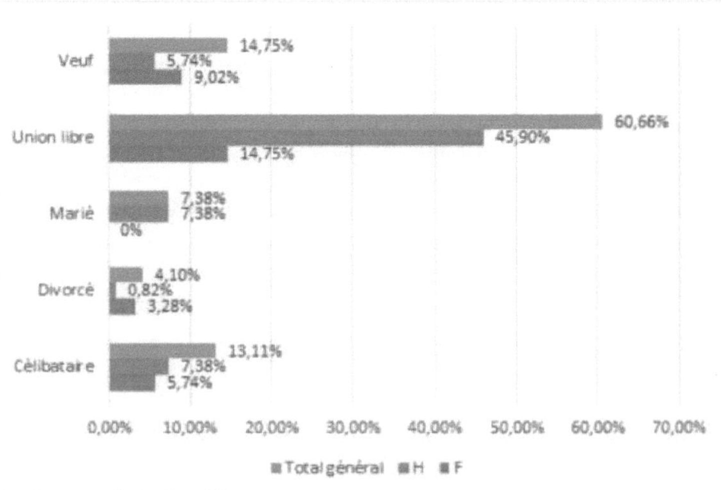

Sources: Kouadio surveys, September 2020

1.2.4 A very high illiteracy rate

Level of education, which covers the following categories: not at school, primary, secondary and higher education. It shows that around three quarters (74.59%) of our target population have not attended school. In addition, only 10.53% of the women had completed secondary education, and none had reached tertiary level. Figure 3 illustrates our analysis.

Figure 3: Level of education

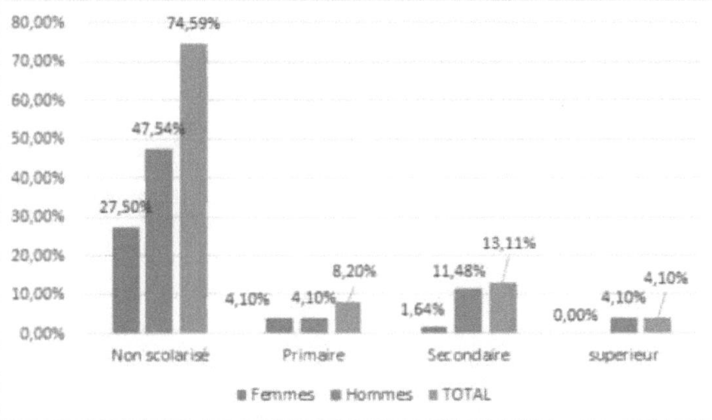

Sources: Kouadio surveys, September 2020

1.2.5. Many children

This is the number of biological children of each returning migrant. The returnees have a total of 844 children, i.e. an average of 6.92 children per actor. Only eight migrants have no children. Of the 16 returnees, 6 are in this mode. Table 3 gives us information on the number of children of each migrant.

Table 3: Distribution of migrant children

Number of children per migrant	Number of migrants	Total children
0	8	0
1	3	3
2	9	18
3	12	36
4	11	44
5	11	55
6	16	96
7	8	56
8	6	48
9	6	54
10	9	90
11	3	33
12	6	72
13	5	65
16	3	48
17	1	17

18	1	18
19	1	19
22	1	22
25	2	50
Grand total	**122**	**844**

Sources: Kouadio surveys, September 2020

1.2.6 A high school enrolment rate for children

Several families had sent their children to school. In fact, 700 of the 844 children of parents surveyed had attended school. We therefore have a school enrolment rate of 82.94% for children born to migrant parents in the Boli sub-prefecture. However, it should be noted that 16 returnees did not send their children to school. Table 4 clarifies the data. There are several possible reasons for this state of affairs.

Table 4: Number of children of returnees attending school

Number of children in school	Number of migrants	Sum of Number of children attending school
0	16	0
1	6	6
2	12	24
3	11	33
4	10	40
5	13	65
6	14	84
7	3	21
8	7	56
9	6	54
10	7	70
11	4	44
12	5	60
13	2	26
15	1	15
16	1	16
19	1	19
22	2	44
23	1	23
Grand total	**106**	**700**

Sources: Kouadio surveys, September 2020

1.2.7 Reasons for children dropping out or not attending school

The children in our target population have a school attendance rate of 82.94%. Some have dropped out of school. The reasons for this situation and for children not attending school vary. According to figure 4, they are, in order of importance, poverty, dropping out, young age (under 6), school failure and the 2002 crisis.

Figure 4: Reasons why children of returnees do not attend school or do not go to school

30,00%

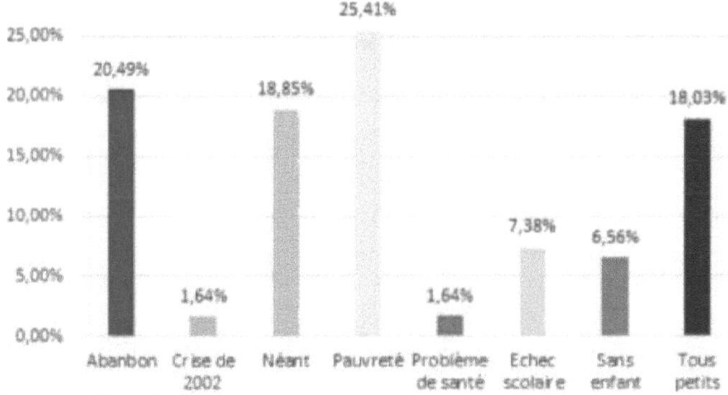

Sources: Kouadio surveys, September 2020

1.2.8 Many working children of migrants

Returnees have several working children. A total of 265 children, born to 73 of the respondents, work. This means that 31.39% of children born to returning migrants are working. The advanced age of the migrants (48.36% are aged 60 and over) and their children's low school attendance (82.94%) are factors that contributed to this rate. Modes 2, 3 and 1 are the most numerous, with 18, 17 and 14 returnees respectively (see table 5).

Table 5: Children of returning migrants who work

Number of working children	number of migrants	Total working children
0	0	0
1	14	14
2	18	36
3	17	51
4	6	24
5	5	25
6	6	36
7	2	14
9	1	9
12	1	12
13	1	13
15	1	15
16	1	16
Grand total	**73**	**265**

Sources: Kouadio surveys, September 2020

1.2.9 Many dependants

Dependants are individuals whose daily survival needs are taken care of by the returnee. They are numerous in the families of our target population. There are 869 people cared for by returnees, i.e. an average of 7.12 children per returnee. It should be noted that 11 returnees have no dependents, but some have dependents of up to 28 individuals. Table 6 illustrates our analysis.

Table 6: People cared for by returning migrants

Number of dependants per migrant	Number of migrants	Sum of Number of dependants
0	11	0
1	5	5
2	7	14
3	12	36
4	8	32
5	15	75
6	14	84
7	6	42
8	4	32
9	4	36
10	7	70
11	4	44
12	7	84
13	2	26
14	1	14
15	8	120
17	1	17
18	1	18
20	1	20
22	1	22
25	2	50
28	1	28
Grand total	**122**	**869**

Sources: Kouadio surveys, September 2020

1.2.10 Christianity, the main religion of returnees

Religious beliefs are widely held. However, there were no Buddhists among the respondents. According to the different religious trends shown in Figure 5, Christians are by far the most numerous, with 59.02% of respondents.

In addition, 8.20% of returnees do not practise any religion.

Figure 5: Religious tendencies of returnees

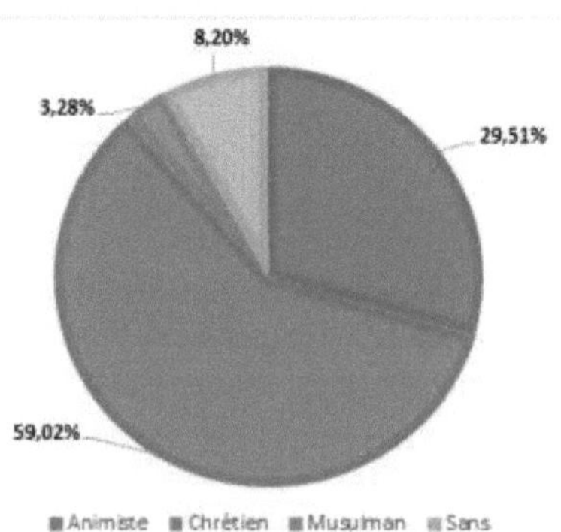

24

Partial conclusion of Chapter I

Returnees are generally over 44 years old, with a sex ratio of 1.97. They are mostly Christian, with an exaggerated female illiteracy rate. They are mostly Christians, with an exaggerated female illiteracy rate. They are monogamous, married by custom and live common-law. They have an average of 6.92 children, over 82% of whom attend school. Around a third of these children work. They have many dependents.

CHAPTER II: SOCIO-ECONOMIC FACTORS, THE CAUSES OF RETURN MIGRATION

People move and reside in a given place for several reasons. Determining the reasons for return migration to the Boli sub-prefecture is important for research into the ways in which these migrants integrate into their localities of residence. In this chapter, which is divided into two main parts, we will first look at the former places of residence and then at the reasons for leaving these places.

2.1 Previous residences before returning to Boli

This involves dividing the returnees firstly according to the area of residence, then according to the number of regions, districts or countries of residence, then according to the last region or district of residence and finally according to contacts with these former places.

2.1.1 Residence areas

The areas of residence are the different zones in which the migrants lived during their emigration from the sub-prefecture. They are either the countryside, an urban area or both. Our survey revealed that return migrants resided in these different zones mentioned above (figure N°6).

Figure 6: Breakdown of players by former area of residence and gender

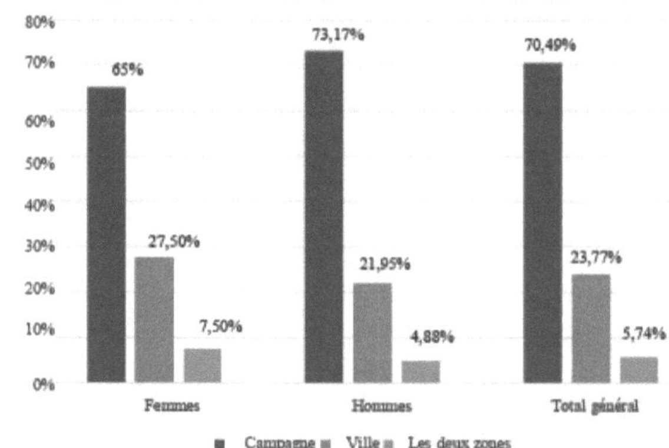

Sources: Kouadio surveys, September 2020

This graph shows the distribution of migrants in their former areas of residence and by gender. It shows that the countryside was the area of residence of the vast majority of migrants (70.49%), with a slight preponderance of men (73.17% for men against 65% for women). After the countryside, those who lived only in the city followed with a rate of 23.77%, before finishing with those who had lived in both areas, who accounted for only 5.74%.

2.1.2 The number of regions, autonomous districts of residence

Emigration from the Boli sub-prefecture took place both nationally and internationally. This section shows the number of different regions and autonomous districts in which the returning migrants resided. This table shows that these migrants travelled through several regions or autonomous districts before returning. For each, the number varies from 1 to 10. Those who lived in just one region make up the largest group, at 40.16% of the total. Only 3 people had visited 8, 9 and 10 different regions and autonomous districts respectively. Table 7 details these routes.

Table 7: The number of regions and autonomous districts covered by each migrant

Number of different regions and autonomous districts	Returning migrants	Frequency (%)
1	49	40,16
2	35	28,69
3	13	10,66
4	9	7,38
5	5	4,10
6	6	4,92
7	2	1,64
8	1	0,82
9	1	0,82
10	1	0,82
Grand total	**122**	**100**

Sources: Kouadio surveys, September 2020

2.1.3 Number of countries covered

In terms of international emigration, our questionnaire shows that very few returnees have crossed the borders of Côte d'Ivoire to establish a residence. In fact, 93.44% of returnees have never lived outside Côte d'Ivoire. 0.82% of returnees have lived in three different countries, 2.46% have lived in two different countries, while 3.28% have each taken up residence in a different country. Table 8 summarises the number of countries of residence.

Table 8: Number of resident countries

Number of different countries	Returning migrants	Frequency (%)
0	114	93,44
1	4	3,28
2	3	2,46
3	1	0,82
Grand total	**122**	**100**

Sources: Kouadio surveys, September 2020

2.1.4 Last or district of residence

These are the last autonomous districts or regions in which the respondents resided before returning to Boli. Table 9 shows this list in descending order to indicate the regions most favoured by returnees during their emigration.

Table 9: Breakdown of returnees by place of residence

Order number	Autonomous resident regions or districts	Number	Representation rate (%)
1	Haut-Sassandra	27	22,13
2	Nawa	17	13,93
3	Marahoué	11	9,02
4	Guemon	9	7,38

5	Abidjan	8	6,56
6	Lôh-Djiboua	7	5,74
7	Yamoussoukro	7	5,74
8	Agneby-Tiassa	6	4,92
9	Cavally	6	4,92
10	Gboklé	4	3,28
11	San-pédro	4	3,28
12	Bas-Sassandra	3	2,46
13	Gbêkê	2	1,64
14	Large bridges	2	1,64
15	N'zi	2	1,64
16	Gôh	1	0,82
17	Gontougo	1	0,82
18	Indénié-Djuablin	1	0,82
19	Moronou	1	0,82
20	Poro	1	0,82
21	Tonkpi	1	0,82
22	Worodougou	1	0,82
Grand total		**122**	**100**

Sources: Kouadio surveys, September 2020

This table shows that the returnees resided in at least 22 different regions and autonomous districts. We note that 15 of the 22 administrative regions that served as last residences are forest regions suitable for farming. These regions are Haut-Sassandra, Marahoué, Nawa, Guémon, Lôh-Djiboua, TAgneby-Tiassa, Cavally, Gboklé, San Péddro, Bas Sassandra, Grands ponts, Indénié-Djablin, Gôh, Moronou and Tonkpi, accounting for 81.97%. The regions most in demand are the coffee and cocoa pioneer fronts of Haut-Sassandra, Nawa and Marahoué. These were the last places of residence for 22.13%, 13.93% and 9.02% of players respectively.

2.1.5 Main activities carried out

The main activities carried out are those that occupied most of the actors' time during their stay outside the Boli sub-prefecture. Figure 7 shows the representation rate of the different activities, which are agriculture, liberal activities (trade, catering and mechanics), salaried activities in the public and private sectors, studies and apprenticeships, and housework. The graph shows that the vast majority of respondents were engaged in farming (67.21%) when they emigrated from Boli.

Figure 7: Breakdown of stakeholders' main activities during emigration

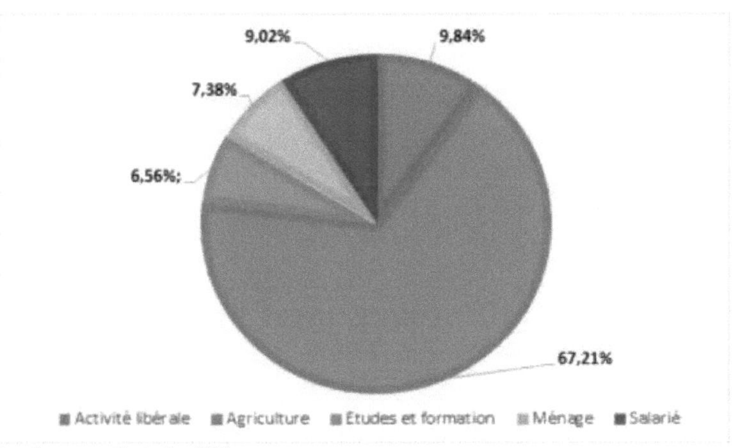

Sources: Kouadio surveys, September 2020

2.1.6 Duration of emigration

This is the amount of time spent outside the Boli sub-prefecture by returning migrants. They have spent several years away from their places of origin. Our investigations and surveys revealed that the time spent emigrating varies from 2 to 35 years. The time spent is divided into five periods:

> [0-5 years [: 4.10% of returnees
> [5-10 years [: 7.68% of returnees
> [10-15 years [: 5.74% of returnees
> [15-20 years old [: 4.10% of returnees
> [20 years and over [: 78.69% of returnees

It is clear that the vast majority of migrants (78.69%) have spent more than
20 years in their areas of origin.

2.1.7 Contacts with former residences

The aim here is to gauge the links that returnees have maintained with their former homes. The contacts are either written or oral communications or regular visits to these places. Figure 8 shows the rate of contact maintained between the players and their previous residences before their return to Boli, which amounts to 50.82%. It can be seen that barely half of the actors (50.82%) maintained contact with their former residences.

Figure 8: Maintaining contact between returnees and former residents

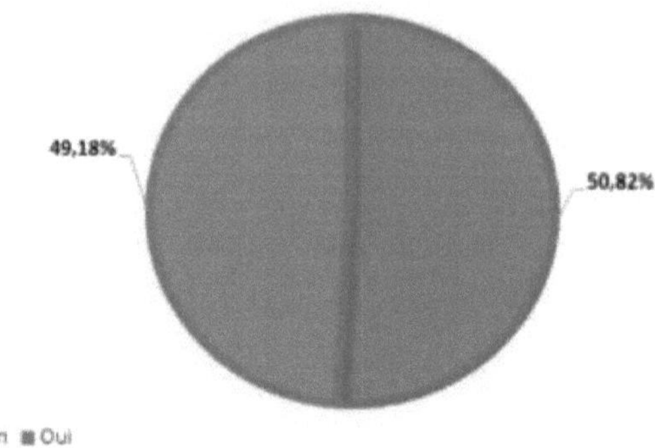

Sources: Kouadio surveys, September 2020.

2.1.8 Gains from emigration

When asked what they had achieved with their emigration income, around two-thirds (63.93%) had not achieved anything while emigrating.

Figure 9: The benefits of emigration

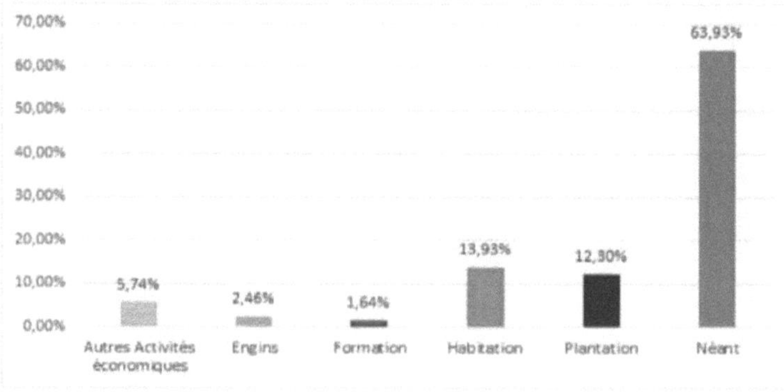

Sources: Kouadio surveys, September 2020

2.1.9 Departure factors from previous residences before returning to Boli

At one point in their lives, our target population emigrated from their place of origin for several years. Today, they are returning to their homeland for a variety of reasons. When asked what factor motivated each respondent to leave their former home, several answers were given. We have divided them into three groups: economic factors with 36.89% (life had become difficult, the need to invest in the village, loss of employment, lack of income), politico-military factors with 20.49% (land conflicts, the 2002 and 2011 crises) and social factors accounting for 42.62% (management of the family in the village, management of the village, death of a relative, neighbourhood problems, pregnancy, marriage, divorce, failure at school, end of training, old age, retirement). Social factors dominate (Figure 10).

Figure 10: Breakdown of factors encouraging people to leave their old homes

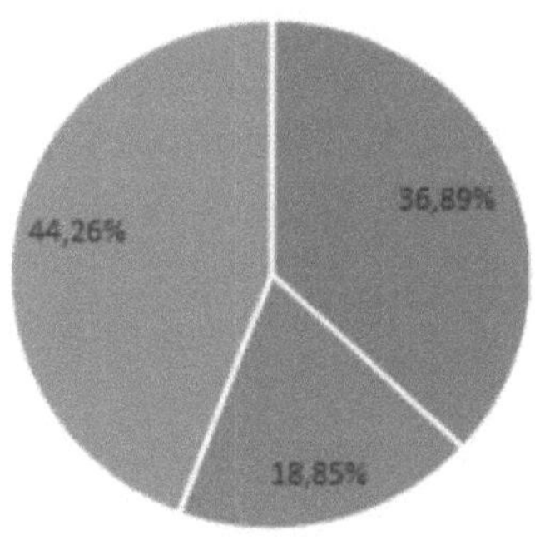

■ Economic ■ Military ■ Social
Sources: Kouadio surveys, September 2020

2.2 Factors relating to the Boli sub-prefecture

2.2.1 Determinants of Boli's departure

These were the reasons that prompted people to leave the Boli sub-prefecture. The vast majority of respondents (69.67%) cited economic factors, in particular logging and the search for employment, as the reason for leaving. Social reasons were also mentioned by 30.33% of respondents. Figure 11 shows the different trends.

Figure 11: Breakdown of reasons for leaving the village.

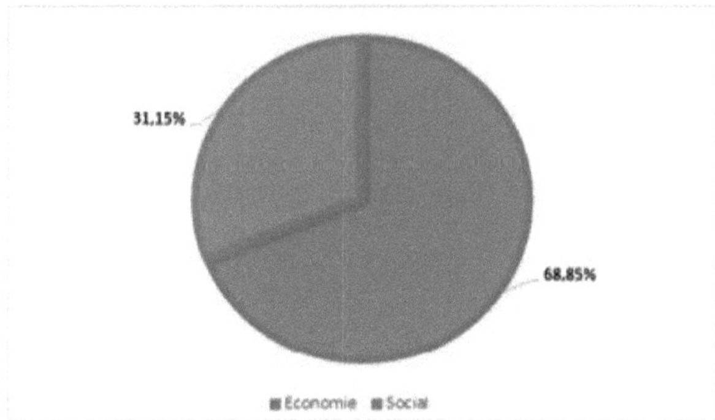

Sources: Kouadio surveys, September 2020

2.2.2 A return to the village accentuated over time

The questionnaire asked about the date of relocation to the village. The answers are grouped according to the following brackets: [0-5 years [, 5-10 years [, 10-15 years [, 15-20 years [and 20 years and more [. Figure 12 shows that the trend towards return migration has increased over the last 10 years (64.75%), especially since 2015, when 39.34% of return migrants were actors. This 10-year period coincides with the post-electoral crisis of 2010-2011. Note that the start of the last 10 years (2011) coincides with the post-electoral crisis of 20102011. There was also a slight increase in the number of arrivals between the ages of [15 - 20], due to the military-political crisis of 2002.

Figure 12: Breakdown of dates of relocation to the village by age group

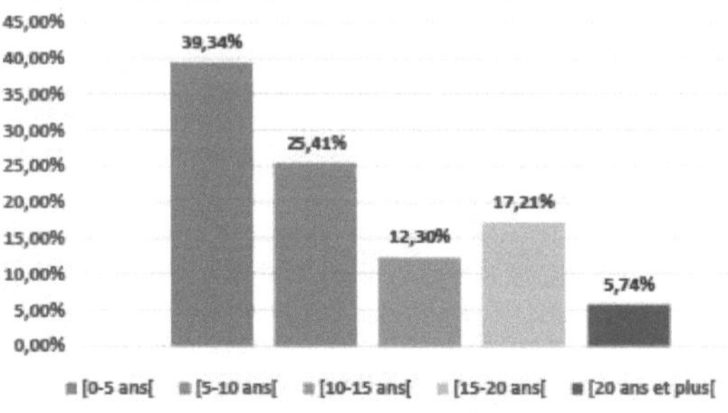

[0-5 years[■ [5-10 years[■ [10-15 years[■ [15-20 years[■ [20 years and more[

Source: Kouadio surveys, September 2020

2.2.3 Reasons for relocating to the village according to age

These are the factors that favoured resettlement in the village, hence the choice to no longer reside in the previous residence. Admittedly, people find reasons they consider sufficient for leaving their previous places of residence, but instead of moving to other administrative districts in Côte d'Ivoire, they preferred to return to the village. Several factors motivated their choice to return to the Boli sub-prefecture. The interviews and questionnaire revealed two main factors: the social factor (rest, family management, village management, security, marriage, retirement, home village) and the economic factor (investing in the village, lower cost of living, availability of land, farming). Figure 13 illustrates the distribution of these factors. Moreover, in general, and contrary to the initial results, social factors account for two-thirds of the total. It should also be noted that people aged 45 and over are the most concerned by the social factor, while younger people have settled mainly for economic reasons.

Figure 13: Breakdown of reasons for resettlement in the Boli sub-prefecture according to age

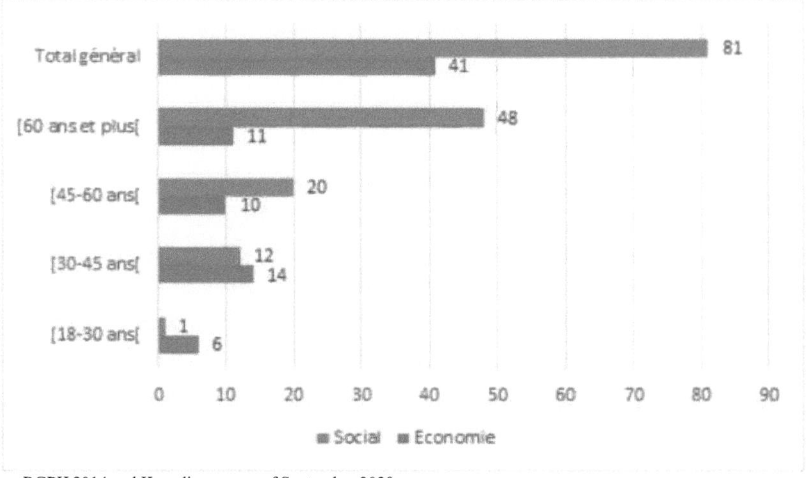

Source: RGPH 2014 and Kouadio surveys of September 2020

Partial conclusion of chapter II

Most of the returnees to the Boli sub-prefecture had been farming for more than 20 years in the forested areas of Côte d'Ivoire. With a completion rate of less than 34%, our target population intensified their return to their homelands from the 2010 post-election crisis onwards for military, economic and above all social reasons. The most migrant people aged 45 and over have returned to the village for economic reasons, while the youngest are there for economic reasons.

CHAPTER III: PEACEFUL COEXISTENCE AND DEVELOPMENT
ECONOMIC ACTIVITIES OF RETURNING MIGRANTS

3.1 Social integration through peaceful coexistence

The social integration of returning migrants in the administrative district of Boli takes into account the residence of other family members, i.e. children and spouses, and relations with other members of the community.

3.1.1 The residences of other family members

3.1.1.1 Children

The children of the returnees either live with them or outside the sub-prefecture of Boli, with some living with them and others outside Boli. According to our investigations, the migrants whose children live with them and others live outside the administrative district are the most numerous. These investigations are summarised in figure 14, which shows that these parents represent 64.91% of all those surveyed. It can therefore be estimated that around two-thirds of the returnees have some of their children living with them and some living outside Boli. A quarter of the actors (25.44%) have their children living far away from them. Only 9.65% surveyed had all their children living under the same roof as them.

Figure 14: Residence statistics for children of returning migrants in relation to their parents

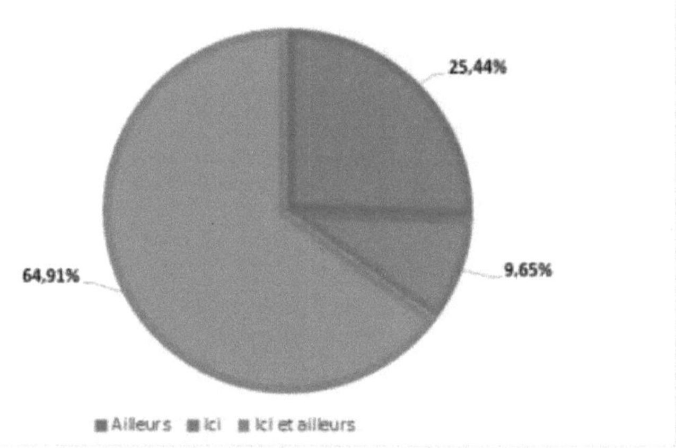

Source: Kouadio surveys, September 2020

3.1.1.2 Spouses

Like the children, the spouses of our target population live either with them, or elsewhere, and either partly with them and partly outside the sub-prefecture of Boli for those who are polygamous. However, in contrast to the results for children, the statistics in Figure 15 show that just over three quarters (77.65%) of respondents with one or more spouses live with them. 15.29% of these respondents have their spouses living elsewhere, while only 7.06% have some of their spouses living with them and some of their spouses living outside Boli. This can be explained by the fact that the spouse or spouses who live outside Boli either carry out an activity that cannot be practised in Boli, or they have remained at the previous residence to manage the husband's work.

Figure 15: Statistics on spouses' place of residence

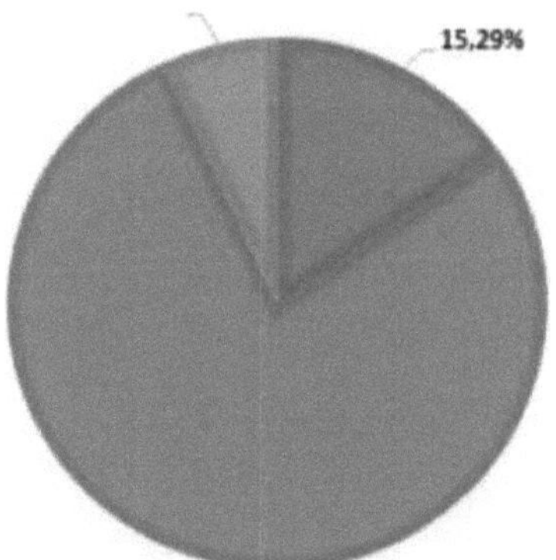

Source: Kouadio surveys, September 2020

3.1.2 Living well with other members of the community

The other members of the community mentioned here are the neighbours of the village or the place where the activity is carried out and the entire population of the sub-prefecture.

3.1.2.1 Neighbours

When asked whether the respondent had ever had a large-scale dispute with a neighbour that required mediation by an authority, whether customary, administrative, police or judicial, with a neighbour in the village or neighbourhood, or with neighbours in their place of work, only 3 people out of 122, or 2.46%, said that they had been confronted with such situations. The sub-prefect and all the village chiefs interviewed also stated that people lived peacefully with the local population. Figure 16 illustrates this. These altercations are linked to politics, a love rivalry and the demarcation of arable land.

Figure 16: Disputes with neighbours

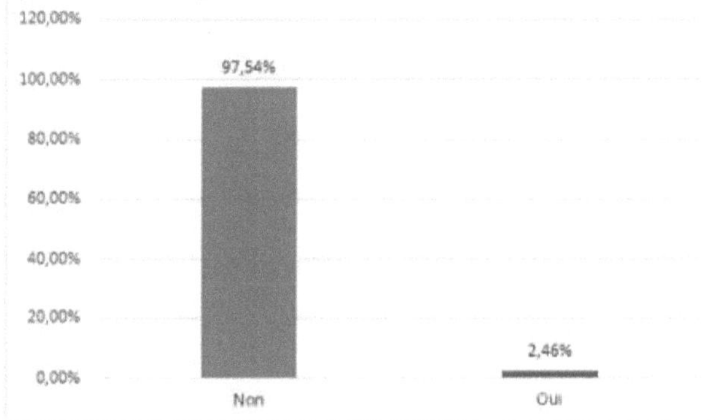

Source: Kouadio surveys, September 2020

3.1.1.1 Level of satisfaction with dispute resolution

It should be noted that our respondents all testified that the village or neighbourhood chief (for those living in Boli) is the only person who settles disputes in their locality. Their levels of satisfaction varied when it came to judging the dispute settlement procedure and the resulting verdict. They ranged from total satisfaction to dissatisfaction to occasional satisfaction. With a frequency of 91.80%, total satisfaction is by far the most widely held point of view. Half-satisfied followed at 6.56%. As for those dissatisfied at any time, they represent only 1.64% of returnees. Figure 17 illustrates the level of satisfaction.

Figure 17: Level of satisfaction with conflict resolution

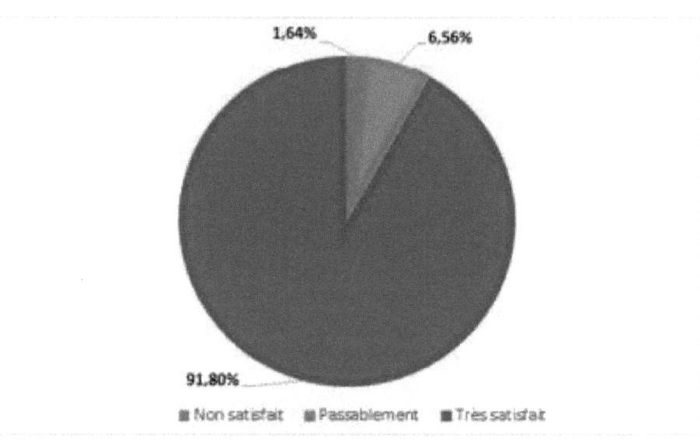

Source: Kouadio surveys, September 2020

3.2. Economic integration through economic activities and achievements
3.2.1 A wide range of activities
3.2.1.1 Main activities

When asked about the various economic activities carried out by the returnees, several different answers were given. These included hairdressing, commerce, dressmaking, export crops, food crops and market gardening, livestock farming, mechanics, attiéké production, charcoal production and catering. Religious figures and housewives were also present. Photo 2 highlights some of the activities carried out by returning migrants. Photo A: Cashew nut plantation; Photo B: Yam field; Photo C: Cassava field; Photo D: Attike factory. From these responses, we note the preponderance of agricultural activities (85.24%). This agriculture is ultra-dominated by food crops such as yams, cassava and sometimes rice. Export crops are mainly cashew nuts and a few oil palm plantations. After agricultural activities, trade remains one of the main activities of those involved, even though only 4.92% of migrants mainly engage in it.

Photo 2: Some of the main economic activities of returning migrants

Photo C   Photo D

Source: Kouadio surveys, September 2020

3.2.1.2 Ancillary activities

Returnees engage in secondary economic activity. These activities occupy more than 61% of this target population. Agriculture plays a major role, as it is practised by more than 54% of the returnees, as shown in table 10.

Table 10: Breakdown of secondary economic activities of returnees

Related activities	Number of migrants	Representation rate (%)
brickmaker	1	0,82
Trade	3	2,46
Sewing	1	0,82
Export crops	38	31,15
Food crops	28	22,95

Breeding	1	0,82
None	50	40,98
Total	**122**	**100**

Source: Kouadio surveys, September 2020

3.2.1.3 The most profitable activities

When people engage in two income-generating activities, it is clear that one of them is more profitable than the other. For this reason, over 80% of returnees derive more profit from agricultural economic activities. Among these agricultural activities, food crops come out on top for 56.56% of returnees. They justify this by the fact that cashew plantations are still young, given the fact that most of them are new arrivals. In fact, 39.64% of the players arrived less than 5 years ago. Following economic activities, trade ranks second with 6.56% of stakeholders. Table 11 gives us the representation rates of the activities that bring in the most income among the returnees.

Table 11: Returnees' most profitable activities

Related activities	Number of migrants	Representation rate (%)
Hairstyle	1	0,82
Trade	8	6,56
Sewing	1	0,82
Export crops	28	22,95
Marshland crops	1	0,82
Food crops	69	56,56
Breeding	2	1,64
Mechanical engineering	1	0,82
les Ménage	1	0,82
None	5	4,10
Prayer	1	0,82
Production of attiéké	1	0,82
Coal production	1	0,82
Catering	2	1,64
Grand total	**122**	**100**

Source: Kouadio surveys, September 2020

3.2.1.4 Problems encountered in carrying out activities

Like other workers in society, most returning migrants encounter problems in carrying out their activities. Only 41.62% of them have no problems. These problems are more of a financial nature. The other difficulties, in order of importance, are: damage from animals, deteriorating health, climatic conditions, lack of labour, few customers, neighbours, lack of arable land and poor infrastructure. Figure 18 details these difficulties.

Figure 18: Difficulties encountered by returnees in carrying out their activities

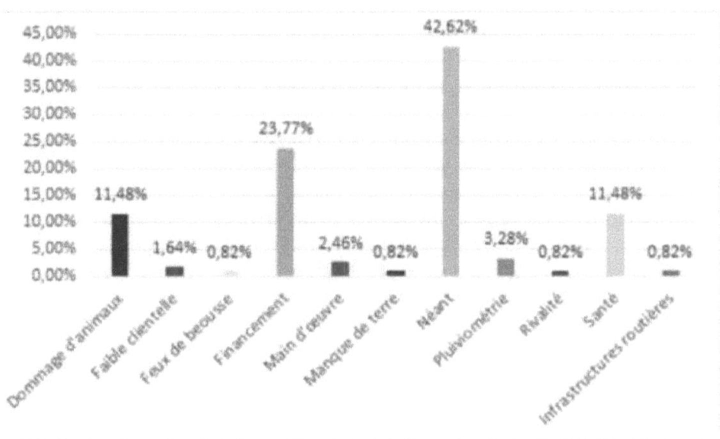

Source: Kouadio surveys, September 2020

3.2.2 Activity space

3.2.2.1 Access to production space

Modes of access to production space refer to the ways in which the returned migrant has gained access to his or her production site. In the sub-prefecture of Boli, our investigations revealed three modalities: family property for 92.62% of the respondents in our target population, leases for 4.92% of them, and gifts for 2.46% of the actors. Figure 19 illustrates our analysis.

Figure 19: Access to production capital

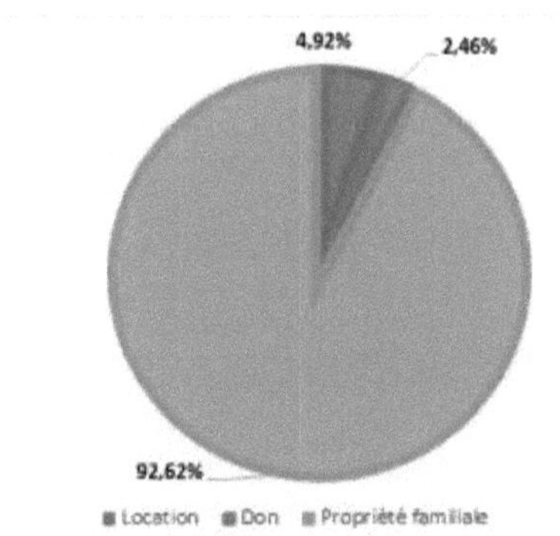

The information channel is the source of information on the availability of production capital for the business. According to our investigations, 4 sources were put forward. First, the parents (86.89%), then the spouse (7.38%), then the migrant himself (3.28%) and finally another source (2.46%) (Figure 20).

Figure 20: Information channel for access to production capital

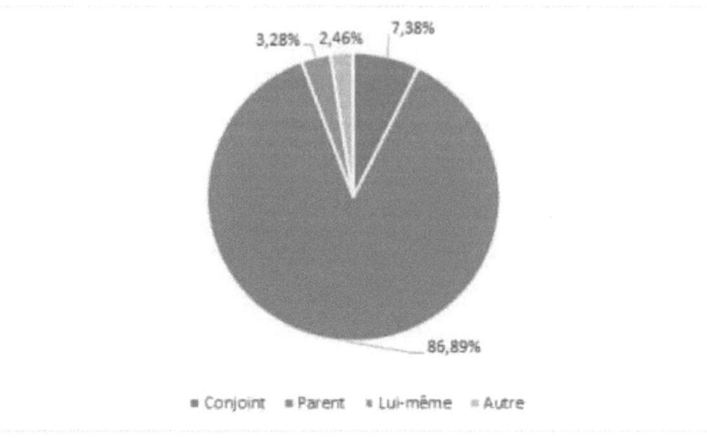

Source: Kouadio surveys, September 2020

3.2.2.3 The time taken by fallowing

The time taken by the fallow is structured by bracket. Fallows of 20 years or more are the most widely used. They are used by 40.75% of migrant farmers. Following them, 28.57% of the stakeholders cultivate those with a resting time ranging from 10 to less than 20 years. Fallow land with a resting period of between 5 and less than 10 years is cultivated by 25.89% of farmers. Only a small portion that is less than 5 years old is used as arable land by 1.79% of return migrants who practise agriculture (figure N°21). People who do not cultivate the land are therefore unable to give time to fallowing.

Figure 21: Time taken by the fallows used

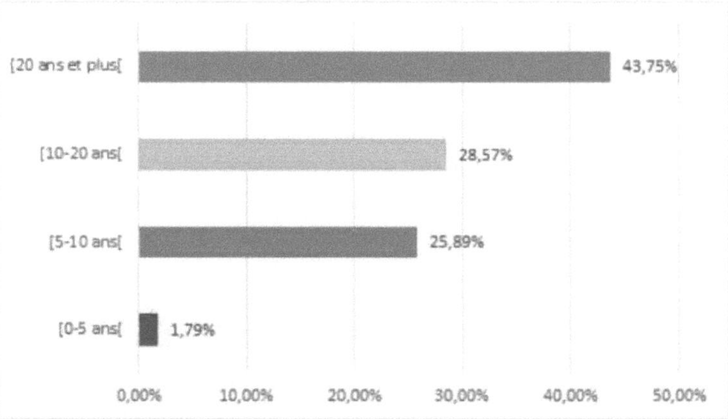

Source: Kouadio surveys, September 2020

3.2.2.4 Soil quality

When asked about the quality of the farmland, 73.77% of the stakeholders found the soil to be of very good quality. However, 19.67% of the returnees thought that the quality of their soil was fair, while 6.56% thought it was poor. Figure 22 is a translation. Our survey showed that cultivated soils are suitable for farming. This is the result of the long time taken by cultivated fallows. Indeed, 72.32% of cultivated fallows are 10 years old or more.

40

Figure 22: Level of satisfaction with soil quality

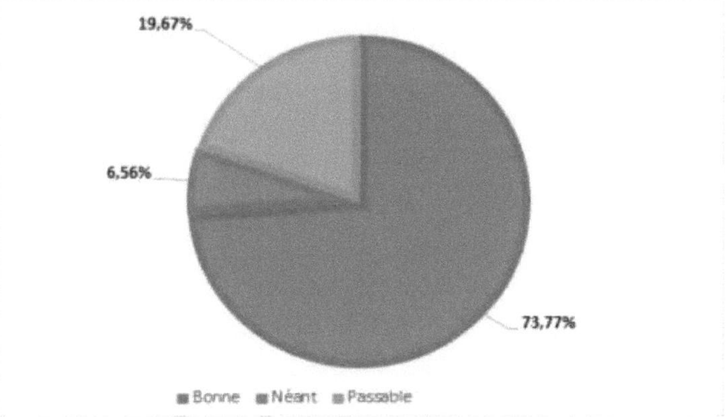

Source: Kouadio surveys, September 2020

3.2.3. A high completion rate
3.2.3.1 Achievements

The returnees to the Boli administrative district have a number of socio-economic projects to their credit, including housing, income-generating activities such as plantations, fields, farms and various other self-employed activities. Photo 3 shows some of these projects. Overall, they have a completion rate of 65.57%. It should be noted, however, that 31.86% of respondents have projects in the agricultural sector (Figure 23).

Photo N°3 : a home and cashew nut plantation belonging to returning migrants

Source: Kouadio surveys, September 2020

Figure 23: Returnees' achievements.

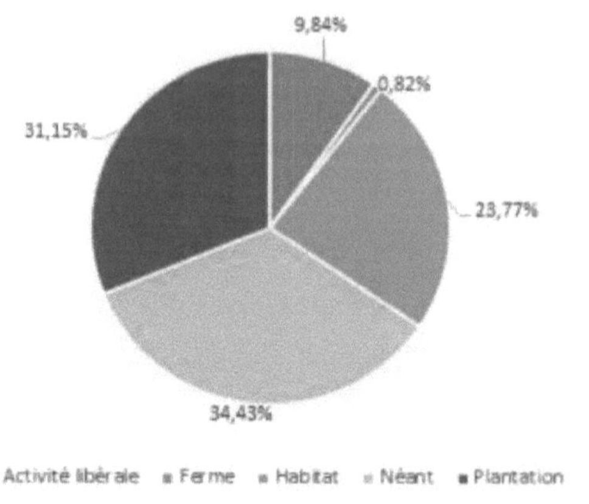

Source: Kouadio surveys, September 2020

3.2.3.2 Achievements by gender
The batch of returnees includes both sexes, with men accounting for 67.21% of the total. Nevertheless, an analysis of the completion rates by sector of activity and overall completion shows a high proportion of women in self-employment. They occupy 58.33% of self-employed jobs and are absent from agro-pastoral activities. They are also less present in housing construction (10.34%) and plantations (21.05%).

Figure 24: Achievements by gender

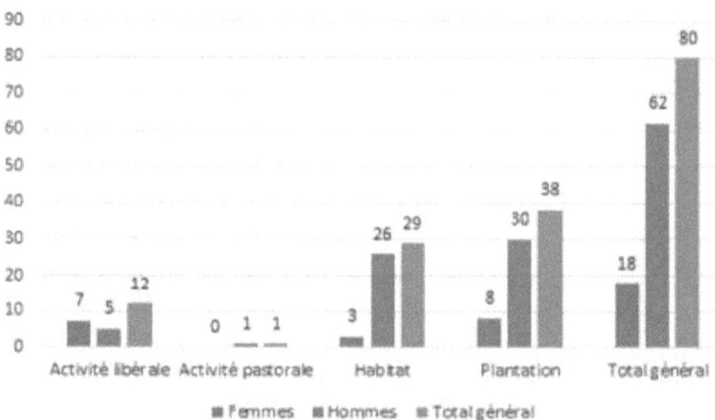

Source: Kouadio surveys, September 2020

Overall, women account for 22.50% of the achievements of returnees in the Boli sub-prefecture, whereas 32.79% of returnees are women. We can therefore conclude that female returnees achieve less than their male counterparts. This low rate of female investment is due to the high proportion of single women and widows among them, which reduces their financial power. Figure 24 provides details.

3.2.3.3 Achievements by age group
Analysis of the migrants' achievements after their return to the Boli sub-prefecture shows an achievement rate of 65.57%. Analysis of the relationship between age and achievements shows that the oldest migrants achieve more than their younger counterparts. In fact, the

completion rates for the age groups [60 years and over [, [45 - 60 years [, [30 - 45 years [and [15 - 30 years [are 67.80%, 66.67%, 61.54% and 57.14% respectively. As for the youngest, those aged [18-30] had a completion rate of 57.14%. This low rate among the youngest can be explained by the fact that they are just starting to invest. Figure 25 illustrates our analysis.

Figure 25: Rate of return migration in each age group

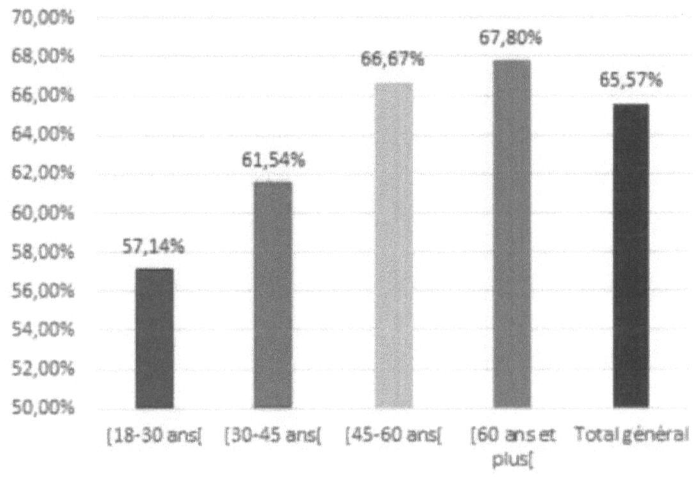

Source: Kouadio surveys, September 2020

3.2.3.4 Return date and completion rate

An analysis of what has been achieved since their arrival shows that all age groups of returnees have an achievement rate of over 50%. However, those who returned more than 10 years ago have a completion rate of over 64%, while their counterparts who returned less than 10 years ago have a completion rate of 57.14%. This difference in realisation rate is explained by the fact that the first arrivals were able to take advantage of their investments, including cashew nut plantations, which is not the case for the new arrivals who have only just started their investments. Figure 26 shows the details of this analysis.

Figure 26: Completion rate for return migrants

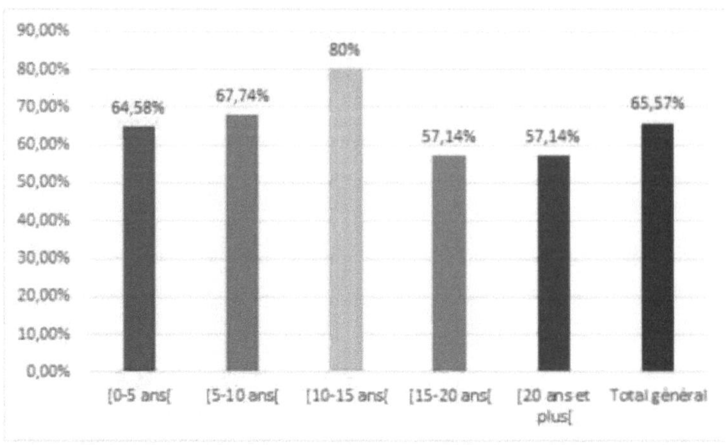

Source: Kouadio surveys, September 2020

Partial conclusion of chapter III

From a social point of view, some of the returnees to the Boli sub-prefecture live with their spouses and some of their children live with them, while others live outside the sub-prefecture. A very large majority have never been in conflict with a neighbour and are in tune with the method of conflict resolution.

From an economic point of view, they practise more subsistence farming on family land, even though they encounter difficulties related to lack of funding, damage from wild animals and health. Despite all these difficulties, they have an achievement rate of 65.57%.

Most of the returnees to the Boli sub-prefecture have been farming for more than 20 years in the forested areas of Côte d'Ivoire. They intensified their return to their lands of origin after the post-electoral crisis of 2010 for military, economic and above all social reasons. Most of the returnees are aged 60 or over, and the majority are male, illiterate, Christian, monogamous and live in common-law relationships. They have an average of 6.92 children, 82% of whom attend school, and many dependents. Their social integration is achieved without difficulty through residence with their respective families and peaceful coexistence with their neighbours throughout the village and sub-prefectural community. Economic integration is largely achieved through agriculture, dominated by food crops grown on family land. They complain of difficulties related to lack of funding, damage caused by wild animals and health. Despite all these difficulties, the returnees to the Boli sub-prefecture have made achievements ranging from the construction of housing to economic activities.

REFERENCES BIBLIOGRAPHIES

AMBROSINI M. (2010), Migrants in the shadows. Causes, dynamics, policies of irregular immigration, article published in the journal "European Journal of International Migration", vol. 26 - n°2

BADIE B (1993), Flux migratoires et relations transnationales, Migrations et relations transnationales Volume 24, number 1

BEAUCHEMIN C. (1999), Emigration urbaine, crise économique et mutations des campagnes en Côte d'Ivoire, article published in the journal "ESPACE, POPULATIONS, SOCIETES", 1999, n°4.

BEAUCHEMIN C. (1999), Emigration urbaine, crise économique et mutations des campagnes en

Côte d'Ivoire, article published in the review "Espace, populations, sociétés" - January 1999

BEAUCHEMIN C. (2000), Le temps du retour ? L'émigration urbaine en Côte d'Ivoire, une étude géographique 406 p.

BEAUCHEMIN C. (2004). Pour une relecture des tendances migratoires internes entre villes et campagnes: une étude comparée Burkina Faso-Côte d'Ivoire. Cahiers québécois de démographie, 33 (2), 167-199.

BROUWEZ C. (2017), International migration: how to act here? Centre Avec.

CERASE F. (1970), "Nostalgia or disenchantment: considerations on return migration" in The Italian experience in the United States edited by S.M. Tomasi and M.H. Engel, Center for migration studies, New York, 1970, pp. 217-239.

DAGNOGO F. ET AL. (2012), "Le chemin de fer Abidjan-Niger : la vocation d'une infrastructure en question", EchoGéo [On line], 20| 2012

DASGUPT A, (1981), Biplab. "Rural-urban migration and rural development" in Why people move edited by Jorge Balan, The Unesco press, Paris, 1981, pp. 43-58.

DAUM C. (2007) Migration, return, non-return and social change in the country of origin In : Petit V. (ed.) Migrations internationales de retour et pays d'origine Nogent-sur-Marne, 2007, CEPED, 157-169. (Meetings)

DIAN B. (200), Aspects géographiques du binôme café cacao dans l'économie ivoirienne,

FLAHAUX M. L. (2009), Les migrations de retour et la réinsertion des sénégalais dans leur pays d'origine, Université catholique de Louvain, 135p

GRIMMEAU J.-? ET AL. (2003), " Tourisme et démographie à l'échelle locale en Belgique ", Espace, populations et sociétés, 2, pp. 263-275

GUBRY P. ET AL. (1996), le retour au village. Une solution à la crise économique au Cameroun? L'harmattan 207p

GUICHARD-CLAUDIC Y. (2001), "Le choix résidentiel de communes rurales bretonnes au moment de la retraite. Des enjeux identitaires diversifiés", Espace, populations et sociétés, 1-2, pp. 139-150

JAUHIAINEN J. S. (2009), "Will the Retiring Baby Boomers Return to Rural Periphery?", Journal of Rural Studies, 25, pp. 25-34, http://dx.doi.org/10.1016/j.jrurstud.2008.05. 001

LESOURD M. (1985), L'exode rural des Baoulé et l'urbanisation de la Côte d'Ivoire. In: Espace, populations, sociétés, 1985-1. Migrations et urbanisation - Migrations and cities. pp. 62-69

LINDSTROM ET AL. (1996), "Economic opportunity in Mexico and return migration from the United States" in Demography, vol. 33, no 3, August 1996, pp. 357-374.

LIPTON M. (1980), "Migration from rural areas of poor countries: the impact on rural productivity and income distribution" in World development, vol. 8, Pergamon Press Ltd, 1980, pp. 1-24.

Mboup, B. (2020) Profils migratoires et insertion socio-économique des migrants de retour au Sénégal, La revue des Sciences Sociales " Kafoudal " N° Spécial Janvier 2020

MIREM (2007), General Report, 162p

UNITED NATIONS (2014), World urbanization Prospects: the 2014 Revision. Population division of the department of economic and social affairs. New York, United Nations. 32 p.

NDIONE B. ET AL. "Diagnostic des projets de réinsertion économique des migrants de retour: étude de cas au Mali (Bamako, Kayes)", European Journal of International Migration [Online], vol. 20 - n°1 | 2004, online since 25 September 2008

NEI, Abidjan: Les nouvelles éditions africaines, 1978, 111 p.

NIEDOMYSL T ET AL. (2011), "Why Return Migrants Return: Survey Evidence on Motives for Internal Return Migration in Sweden", Population, Space and Place, 17, pp. 656-673

OECD (2017), "Capitalising on return migration by making it more attractive and sustainable" in Interrelations between Public Policies, Migration and Development, OECD Publishing, Paris. 283p.

IOM (2013), International Migration, Health and Human Rights, 2013, 68 p

ON-JOOK ET al, "Adaptation in the city and return home: a dynamic approach to urban-to-rural return migration in the Republic of Korea" in Why people Move, edited by Jorge Balàn, The Unesco Press, Paris, 1981, pp. 230-241.

PETIT ROBEERT: https://dictionnaire.lerobert.com/defmition/retour 05/10/2020

PETIT V. (2007), Migrations internationales de retour et pays d'origine, CEPED, Rencontres 208p

PIGUET E. (2013) Les théories des migrations. Synthesis of individual decision-making, article published in the journal "in Revue européenne de migrations internationales" - September 2013

POTVIN D. (2006), Les jeunes adultes migrants de retour, un potentiel pour le développement de leur région d'origine, 309p.

RALLU J. L. (2003) Institut national d'études démographiques, Paris, Démographie:

RGPH 2014: General Census of Population and Housing 2014

SANDERSON J.-P. (2015), Retour des retraités en ville : Mythe ou réalité ? Étude des migrations des 50-69 ans à Bruxelles, REVUE QUETELET/QUETELET JOURNAL Vol. 3, n° 1, October 2015, pp. 51-74

THOMSIN L. (2001), "Les mobilités de la retraite", M. LEGRAND (ed), La retraite : une révolution silencieuse, Éres, Toulouse, pp. 223-242.

VON REICHERT C. (2001), "Returning and New Montana Migrants: Socio-economic and Motivational Differences", Growth and Change, 32, pp. 447-465

WALDORF B. (1995), "Determinants of international return intentions" in Professional Geographer, vol. 47, no 2, 1995, p. 125-136.

WYMAN M. (2001), "Return migration-Old story, New story" in Immigrants&Minorities, vol. 20, no 1, March 2001, pp. 1-18.

ZEKRI B. H. (2007), La migration de retour en Tunisie. Etude du cadre législatif, du contexte socioéconomique et des processus de réinsertion des migrants de retour, Rapport d'analyse MIREM-AR 2007/04.

QUESTIONNAIRE

SOCIO-ECONOMIC INTEGRATION OF RETURNING MIGRANTS IN THE BOLI SUB-PREFECTURE

FICHEN°   DATE: /   /2020

I.   PROFILE OF RETURNEES TO THE SUB-PREFECTURE OF BOLI

1.   Place of residence :   QuartierVillage
2.   Age : [18-30[□ ; [30-45[□ ; [45-60[□ ; [60ans et plus [□
3.   Genre: H □ ; F □
4.   Are you a native ;   Allochthonous □ ; Foreign □
5.   Ethnic group :
6.   Your nationality :
7.   Marital status: Married □ ; Common-law □ ;   Single □ ;   Divorced □ ; Widowed □
8.   Are you? Monogamous □ Polygamous □ Polyandrous □
9.   Your spouse(s) is/are: Aboriginal □; Allochthonous □; Foreign □
10.   Number of children   ,
11.   Number of girls ;
12.   Number of boys
13.   Number of children in school
14.   Number of working children
15.   Number of children not attending school and/or (from) school
16.   Why is this?
17.   Number of dependants
18.   Level of education: Not enrolled □ ; Primary □ ; Secondary □ ; Higher □
19.   How much did you spend on your activities?
20.   Religion: Christian   ; Muslim □ ; Buddhist □ ; Animist □ ; None □

II.   DETERMINANTS OF RETURN MIGRATION IN THE SUB-REGION PREFECTURE OF BOLI

21.   Reasons for leaving the village
22.   When did you leave the Boli sub-prefecture?   [0-5   years [□ ; [5 10 years [□ ; [10-15 years [□ ; [10-20 years [□ ; [20 years and over [□
23.   Were you in the countryside □ ; In the city □ ; Both areas □ ?
24.   How many different regions have you lived in?
25.   How many different countries have you lived in?
26.   Main activity during your emigration:
27.   Ancillary activity carried out during your emigration:
28.   In which region did you last reside?
29.   Reasons for leaving previous residence
30.   Contact with your starting point: Yes □ ; No □
31.   Reasons for relocating to the village?
32.   How long have you been resettled in the Boli sub-prefecture? [0-5 years [□ ; [5-10 years [□ ; [10-15 years [□ ; [15-20 years [□ ; [20 years and more [□
33.   What have you achieved during your emigration?

111.INTEGRATION INTO THE SOCIO-ECONOMIC FABRIC
IN THE SUB-PREFECTURE OF BOLI

34. Place of residence of your spouse(s): Here ; Elsewhere □ ; Here and elsewhere □ ; None
□

35. Your children's place of residence: Here □ ; Elsewhere ; Here and elsewhere □ ; None □

36. Have you had any disputes with your neighbours in the neighbourhood or village? Orii ;
No □

37. If so, why?

38. Have you seen an increase in community problems here? Yes□ ; No□

39. What are the causes?

40. Dispute resolution bodies?

- the community chief □-   the village/neighbourhood chief □-   the sub-prefect □

- Gendarmerie □ - Court Other (specify) :

41. Level of satisfaction with dispute resolution: Very satisfactory □;

FairlyQ; Not satisfactory □

42. Access to production capital :

-      Family property □-   Cash purchase-   Labour □-   Loan □

-      State allocation □-   Rental □-   Donation □-   Other : □

43. Information channels relating to the availability of production capital: a parent □
your spouse □ yourself □ other : □

44. Do you have a "title deed" (a piece of paper)? YesNo □

45. Did you have any problems settling in? YesNo   □

46. If so, which ones?

47. Have you had any disputes with your neighbours at your place of business? Yes No □

48. Why is this?

49. Main activity after your return to the village:

50. Additional activity after your return to the village:

51. What activity brings you the most income each year?

52. What is the surface area of your crops or business premises?

53. How long does the fallow land you cultivate take? : [5-10[years ; [10-20[□ years ;
[20

years and over [□

54. How would you rate the quality of this soil? Poor □ - Fair □ - Good □

55. Problems encountered in carrying out your activities 1 :

2 :

56. Have you improved your residential habitat? Yes ; No □

57. Achievements from your village activities: 1:

2

yes

I want morebooks!

Buy your books fast and straightforward online - at one of world's fastest growing online book stores! Environmentally sound due to Print-on-Demand technologies.

Buy your books online at
www.morebooks.shop

Kaufen Sie Ihre Bücher schnell und unkompliziert online – auf einer der am schnellsten wachsenden Buchhandelsplattformen weltweit! Dank Print-On-Demand umwelt- und ressourcenschonend produzi ert.

Bücher schneller online kaufen
www.morebooks.shop

Printed by Books on Demand GmbH, Norderstedt / Germany